SCIENCE FAIR

HOW TO DO A SUCCESSFUL PROJECT

HUMAN BODY

BY

JOHN M. LAMMERT

SERIES CONSULTANT
DR. JOHN M. LAMMERT
Associate Professor of Biology
Gustavus Adolphus College
St. Peter, Minnesota

ROURKE PUBLICATIONS INC.
Vero Beach, FL 32964
U.S.A.

LIBRARY OF CONGRESS CATALOGUING-IN-PUBLICATION DATA

Lammert, John, 1943–
The human body / by John Lammert
p. cm. — (Science fair)
Summary: Explains the scientific method and suggests a variety of experiments and projects about the construction and function of the human body
ISBN 0-86625-426-9
1. Human physiology—Juvenile literature. 2. Human anatomy—Juvenile literature. 3. Science projects—Juvenile Literature.
[1. Human physiology. 2. Human anatomy. 3. Science projects.]
I. Title II. Series.
QP37.L248 1993
612'.0078—dc20

92-6975
CIP
AC

DESIGNED & PRODUCED BY:
MARK E. AHLSTROM
(The Bookworks)

PHOTOGRAPHY:
Cover–THE IMAGE BANK/Joe Saffold
Text–MARK E. AHLSTROM

The publisher and author accept no responsibility for any harm that may occur as a result of using information contained in this book.

TABLE OF CONTENTS

CHAPTER 1

What Makes A Science Project "Scientific?"

In a good science project, you will be an investigator. You will look at clues provided by the world about you and will make some sense of them by planning and doing an experiment. This is what scientists do when they observe some natural phenomenon and want to investigate it—they use the **scientific method**. In an organized fashion, scientists follow a series of steps that are designed to help them come up with an explanation for something they have observed.

When scientists observe something new, they wonder "How can that be?" The scientists have identified a problem. A scientist then makes an educated guess, or a **hypothesis**, that might explain the observation or solve the problem. It is "educated" because the scientist has some knowledge of the subject matter or reads what other scientists have found out through their experiments about the problem.

Steps Followed in the Scientific Method

- ✔ Make an *observation*.
- ✔ State the problem: What do you want to find out?
- ✔ What is already known about the observation?
- ✔ Develop a *hypothesis*: What do you think is a reasonable explanation for the observation?
- ✔ Design an *experiment* that will provide answers: What materials will be needed and how will they be used?
- ✔ Record *data* or observations: What happened during the experiment?
- ✔ Analyze the results.
- ✔ Come to a *conclusion*: What did you learn? Did your data support your hypothesis? What do your results mean?

Let's look at an example of an observation. From this observation, an hypothesis will be made.

Different foods may have different tastes. A cherry-flavored sucker tastes sweet. A bag of popcorn at the movies tastes salty. A grapefruit eaten at breakfast tastes bitter. A sweet pickle tastes sour. One reason we need the sense of taste is so we can detect safe foods, which usually have a pleasant flavor, and spit out something that might harm us. Poisonous plants tend to taste "yucky." Our tongues have **taste buds** that detect flavored substances dissolved in water. These taste buds can tell the difference between four basic flavors—sweet, salty, bitter, and sour. Taste buds are found within the small bumps you can see on your tongue. They are connected by nerves to the brain. When a taste bud is stimulated by a particular substance, a nerve carries a signal to the brain, which figures out what the substance's flavor is. A good question about taste might be "Where on the tongue do we detect each of these four basic flavors?" You might then make the hypothesis that "Each flavor is tasted on different parts of the tongue."

Different foods have different tastes.

Next, the scientist makes a plan to test the hypothesis. This plan is called the **experimental design**. Materials are chosen and a series of steps are written out. Then the actual work can start.

Here is an experimental design for our hypothesis that there is a different part of the tongue for each of the four basic flavors. In this experiment, solutions of different flavored substances—sweet, salty, bitter, and sour—are dabbed on various places on the tongue. A record is kept of which parts of the tongue taste each flavor.

Sugar solution on a cotton swab is placed on the tongue.

Then the actual work can start. Several people are asked to be **experimental subjects**. They agree to have their tongues "mapped" for places where these flavors are tasted. Solutions of sugar, salt, and vinegar are mixed. An aspirin dissolved in a little water is the "bitter" solution. If any subjects are allergic to aspirin, unsweetened grapefruit juice is a good substitute. A flat toothpick or cotton swab is used to apply small amounts of one solution to the tip, edges, middle, and back of the tongue. The toothpick or swab is rolled around the edge of the glass to remove excess solution. Subjects are not told which taste is being used. After a little bit of a solution is placed on the tongue, the subject says what flavor is tasted. After each application, the subject rinses off the solution from the tongue with a little water. A new toothpick or swab is used for each taste solution.

During the experiment **data** are collected. This means that what you see or what you measure is recorded. In our example, a "taste map" of the tongue is made for each of the four basic classes of flavors. An outline of the tongue is drawn on a piece of paper. Each place on the tongue where a particular taste is detected, a small "o" or other symbol is written down on the taste map. Four taste maps are prepared, one for each basic taste.

The **results** are studied. Here are some things to look at. Are there some places on the tongue where more than one flavor was tasted? Do some of the "taste regions" overlap with each other? Are there some places where nothing could be tasted?

Finally, from these results, **conclusions** are made. These questions are answered by the scientist. What was learned? Did all parts of the tongue taste all flavors? Did different sections taste only one flavor? Were two or more flavors detected in the same area? Was the hypothesis correct, that is, did the data and the hypothesis agree? If it was not correct, then what would be another hypothesis?

The scientist will then use these data to make some **predictions** that will lead to more experiments. When the data from an experiment have been studied, there are often new questions that require answers. A scientist's work is never done. A scientist doesn't stop doing experiments even after winning the Nobel Prize!

CHAPTER 2

Choosing The Human Body As A Topic For Your Project

What makes a good science project? Students with experience will probably tell you that the hardest part is finding a topic. Take plenty of time trying to think of a good topic. This book is written to help you decide what to do. It's important that you are interested in the topic you choose. This means that you will be excited about what you are doing. You usually learn better when you are excited about something. Don't think you must have a difficult project, especially if this is your first. The best projects often focus on something simple.

Many people want to learn more about the human body. It's fun to discover how things work inside of you. We chew and digest food so that our bodies can get the building blocks needed for growth. Our senses, such as vision, knowing which way is up, taste, touch, and hearing, tell our brain what is happening outside the body. Our bodies have special organs to get rid of water and remove harmful things from our blood. We have cells to fight infections caused by nasty germs. Lungs expand to fill with oxygen and deflate to expel carbon dioxide. Our hearts beat to pump the blood that carries oxygen from our lungs to the rest of the body. We inherit traits from our parents and grandparents.

This book will guide you through the basic steps you need to follow so you can complete a successful science project about the human body. In the next few pages you will read about project ideas that focus on the human body. You will not find step-by-step instructions. Instead, the ideas are here to spark your imagination and to turn on your creative powers. That's what makes science fun—imagination and creativity.

There are many investigations you can do about your skin, your eyes, your ears, and your mouth.

What types of projects can you do about the human body?

There are four different types of projects you can choose to do about the human body:

- ☆ You can prepare an *exhibit* that displays a collection or shows what you learned after reading about a topic.
- ☆ You can do a *demonstration* of a lab experiment you found in a book.
- ☆ You can conduct a *survey* in which you observe the activities of many people or gather the opinions of many people.
- ☆ You can do an *investigation* that uses the scientific method to explore a problem.

Sometimes science fair rules discourage the first two types of projects. Teachers or judges may feel that these projects are not scientific enough. However, you might be able to do an exhibit or demonstration *if* you add a little investigation to the project. It's a good idea to have your teacher go over your project plans with you. This will help you find out if your project satisfies your local rules.

☞ *Exhibits*

The body needs to know what is happening outside of it so that it can make the proper inside adjustments. You may hear a rustling noise in the bushes as you walk through the woods at night. You bite a grape and taste its sweet flavor. You are hungry, so you look in the kitchen for a cookie or potato chips. These changes around the body are detected by special structures called **sense organs**, which serve as "windows to the world." In these sense organs are **receptors** that *receive* information from outside the body. This information is called a **stimulus**. A stimulus could be light, a sound, pressure, or temperature. There is one type of receptor for each class of stimulus. In a receptor, the stimulus is changed into a signal that is carried by nerves to the brain. Two pieces of information about the stimulus reach the brain:

what was the stimulus (color of light, pitch of sound) and how strong was the stimulus (bright or dim, loud or soft). You might create an exhibit that presents what you learned about the workings of one of these senses. Ideas for such exhibits include:

- ❖ How the ear hears sound. (What is sound and how is the ear designed to collect sound waves?)
- ❖ Light and the eye. (What is light and how is the eye designed to receive this light?)

Our bodies have organs that perform specific functions. Many times several organs cooperate so that the body gets something it needs for growth. For example, the digestive system includes several organs, each contributing to the breakdown of the food we eat. These organs include the stomach, the pancreas, the small intestine, and the large intestine. The circulatory system uses the heart to pump blood through blood vessels. In the blood is food that body cells need for growth, red blood cells that carry oxygen for other cells as they extract energy from food, and white blood cells and protein substances that protect us from infections. **Hormones** are chemicals made by certain cells that signal other cells to do something. **Insulin** is a hormone released from the pancreas, an organ that lies just below the stomach. Insulin signals cells in the liver and muscles to take up sugar from the blood. Here are some ideas for an exhibit about an organ or organ system:

- ❖ Digestion—What happens to a bite of pizza?
- ❖ Nutrition—Eating right and living right.
- ❖ The heart—The pump that moves blood.
- ❖ Blood—The fluid of life.
- ❖ The lungs—A breath of life.
- ❖ Skin—Not just a covering.
- ❖ Hormones—The body's messengers.
- ❖ Bones—A skeleton in the closet.

Sometimes something goes wrong in an organ and a disease can occur. The heart needs to be fed constantly by blood that flows in **arteries** around and through the heart. If these blood vessels become clogged, the heart doesn't get needed oxygen and food. Without food, the heart muscle becomes damaged. The result can be a heart attack. Medical scientists think that several things play a role in the slow buildup of blockages in arteries. These include too much fat and cholesterol in the diet, high blood pressure, and smoking cigarettes. Americans eat a lot of fatty foods—hot dogs, hamburgers, french fries, and doughnuts, to name a few. Salt has been found to increase blood pressure for some people. Think of all the foods you eat that are salty. Doctors are now discovering that children are affected by too much fat, cholesterol, and salt. Only a few years ago, they thought that only older people should be concerned about these things. Now, doctors are concerned about children who don't eat right.

Too much of these foods, that you may like to eat, can be harmful to your heart.

Some diseases develop when an organ doesn't make enough of the hormone that it should. For example, when the pancreas doesn't make insulin, the disease **diabetes** occurs. So a person with diabetes must get a shot of insulin every day to help get sugar out of the blood and into body cells.

Here are some ideas for an exhibit that shows what you learned about a disease and what you want to teach others about it.

- ❖ Cholesterol and heart disease
- ❖ High blood pressure
- ❖ Diabetes
- ❖ Smoking is hazardous for your health
- ❖ Teeth and cavities
- ❖ The sun and skin cancer
- ❖ AIDS and the immune system

In preparing your exhibit, you might decide to design and make a *model* that shows the main structure of the **organ** that is responsible for the body activity you are presenting. However, don't use a plastic kit that you bought and assembled. This does not show any creativity by you.

Carefully drawn illustrations are a good idea to include in your exhibit. However, don't tear out pictures from magazines and books. Photocopies of pictures in books should not be used. Any written material that you use in this exhibit should be in *your own words*. Your exhibit will be judged in part on creativity and originality.

☞ *Demonstration*

The library may have some books that describe how to do some experiments that investigate some feature of the human body. Sometimes these are too simple and only take a short time to do. If you decide to do one of these easy projects, it will probably receive a low rating by your teacher or science fair judge.

After you read the next section, put on your thinking cap and see if you can add something to a demonstration that will turn it into an investigation.

☞ *Investigation*

The best type of science project shows an investigation that you carried out. By exploring some feature of the human body in an investigation project, you will learn more about how scientists think and work. You will also learn some important things about the world around you.

Let's review again the questions you need to ask as you plan and carry out a project that uses the scientific method:

- ✔ What do you want to find out? This is the *problem* that you have decided to investigate.
- ✔ What is already known about the problem?
- ✔ What is your experimental design for an experiment that will give you an answer to your problem?
- ✔ What do you think will happen? This is your *hypothesis.*
- ✔ What did you observe happen? These *observations* are written in your lab notebook.
- ✔ Analyze your data.
- ✔ What did you learn from the observations? Did you find that your hypothesis was correct? If not, can you think of another hypothesis?

Here are some suggestions for a question to ask about the human body.

You may not know it, but a clock ticks inside of us. This body clock counts out cycles that have peaks and valleys. We do some things better at certain times of the day. For example, scientists has found that thinking and taking exams is better between 1:00 and 7:00 p.m. than other times of the day. Physical activity, like playing baseball or football is also best at these times. The worst time is during the middle of the night. Our body clock also creates rhythms of temperature, heart beat, and **blood pressure**. So what questions can you ask about the human body clock?

❖ What is the daily pattern for body temperature? Use a mouth (also called oral) thermometer to read your temperature every several hours during the day. Begin when you get up and finish just before you go to bed. When you take your temperature, be sure to keep the conditions around you the same, like air temperature and how much physical activity you just did. Do this for several days. Ask several family members or friends if they will have their temperatures taken also. Compare the patterns of temperature measured over the time when people are awake. You could investigate if the body temperature cycle differs for young and old people or for boys and girls.

❖ What is the daily pattern for heart beats? You can detect the beat of your heart by feeling a *pulse*. When heart muscle contracts, a surge of blood shoots out of the heart. The arteries that carry this blood are elastic and so will expand and then recoil with each heart contraction. Usually the pulse can be felt about two to three centimeters (one inch) below the base of your thumb. Lay your left hand out on a flat surface, wrist up. Lightly put the first two fingers of your

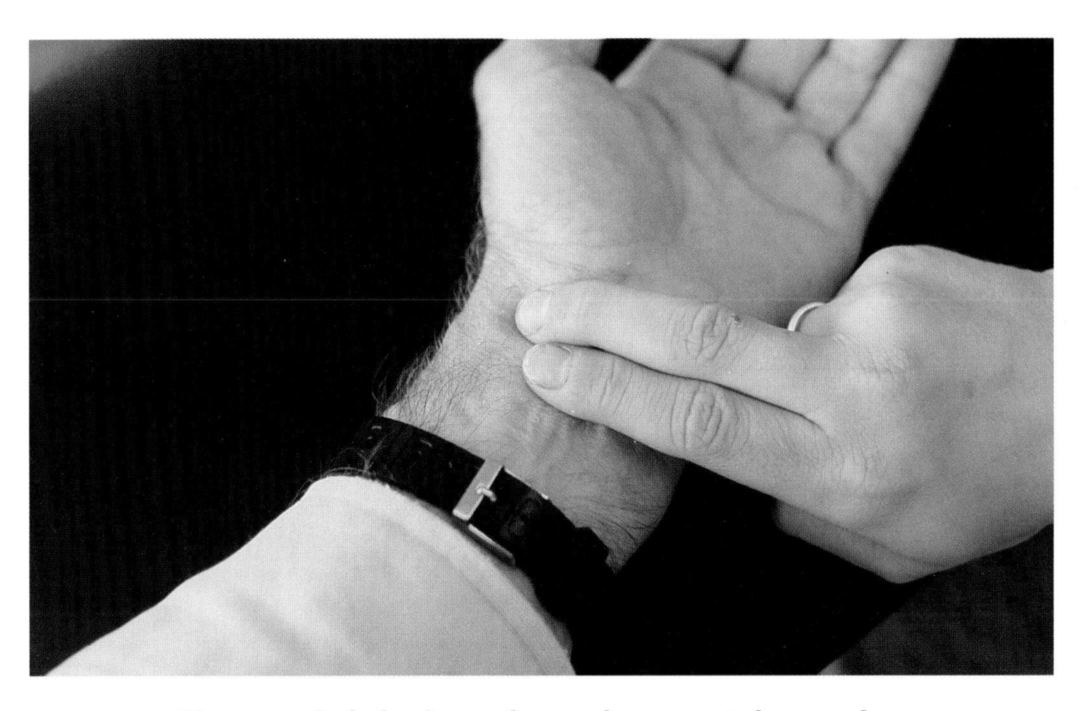

You can feel the heart beat when you take a pulse.

right hand over the pulse area of your left wrist. Move your fingers just a little if you can't feel the pulse. Count the number of times you feel the heart beat for 30 seconds and multiply by two. This gives the number of heart beats a minute. Be sure to keep the conditions around you the same each time you take a pulse. Ask other people if you can use them in your investigation. A question that might make this a better investigation is, "Does the daily pattern of heart beats per minute differ for young and old people or for boys and girls?"

❖ What is the daily pattern for blood pressure? Blood pressure is the pressure of blood against the wall of an artery. When the heart contracts, blood leaves under pressure. Blood pushes against artery walls. This force is called **systolic** (*sis TAHL ik*) **pressure**. When the heart relaxes before the next contraction, the pressure on artery walls is less because no blood is being pumped out of the heart at that moment. This is called **diastolic** (*DIE ah STAHL ik*) **pressure**. Blood pressure is measured with an instrument called a **sphygmomanometer** (*sfig moh MAN ah met er*). You might be able to borrow an electronic sphygmomanometer that is easy to use. It will give two readings—the first is the systolic pressure and the second is the diastolic pressure. Measure your blood pressure throughout the time you are awake. Before you take your blood pressure, sit quietly for two to three minutes. Is there a daily pattern of blood pressure with hills and valleys? Continue to record your blood pressure for several days. Be sure to keep the conditions around you the same each time you take a pulse. Ask other people to participate in your study of blood pressure. Use the data you obtain from these people to see if they all have the same pattern or if age or sex makes a difference.

Blood pressure and heart rate might be influenced by other factors than the time of day. Explore changes in these heart activities that might take place when people are in different body positions, such as lying down or standing on their heads. If there are differences, see if you can explain why. When blood vessels near the skin contract because the body is exposed to cold, the space available for blood to move through becomes less. What will happen to blood pressure and heart rate? Test your hypothesis by immersing one hand of an experimental subject in ice water and measuring this person's blood pressure and heart rate every 30 seconds for two minutes. Certain drugs affect heart activity. Caffeine affects the heart. Measure blood pressure and pulse rate every 5 minutes for an hour after several experimental subjects drink a carbonated beverage that contains caffeine.

Our skin has many nerves that transmit signals to the brain to inform it of changes around the body. **Sensors** at the skin end of these nerves detect these changes. One kind of sensor will detect temperature. Others sense touch, pain, or pressure. The brain uses this information to decide what action to take. For example, if your finger touches something very hot, you quickly pull it away. The brain tells your arm and hand muscles to move the finger from the danger of burning. These signals are important in protecting us.

A question for an investigation about one of these sensors is: "Are the touch receptors equally distributed on the skin?" In this investigation, you could measure how far apart touch receptors are on the different parts of the face, hands, arms, back, legs, and feet. Ask several experimental subjects to agree to participate in your investigation. To measure the distance between touch receptors, first cut out a piece of cardboard that is 2.5 centimeters wide and 6 centimeters long. Then make a small line every 2 millimeters for 5 centimeters along the middle of the cardboard. Make a longer line at

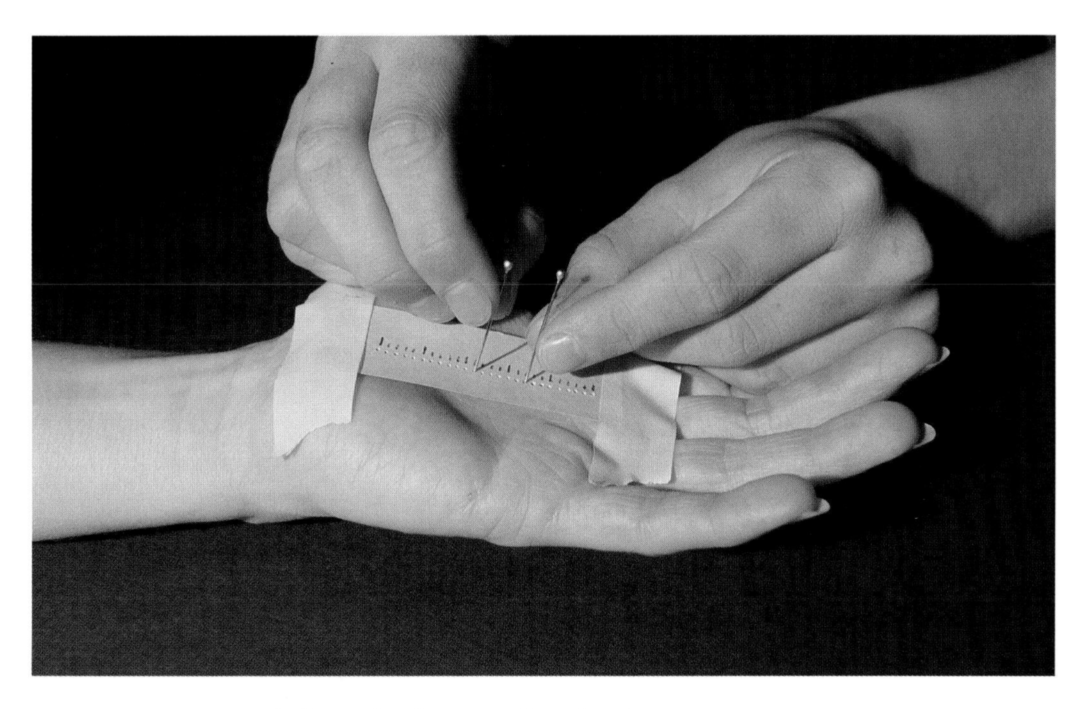

A "touch distance" ruler is easy to use.

each centimeter. Take a small nail and push it through the cardboard to make a hole at each mark. Scrape two needles on concrete to dull their points a little. Your subjects should close their eyes as you carry out the experiment. Put one needle through the first hole and press it *gently* on the skin. Then put the second needle through the second hole and press it *gently* on the skin. If the subject can feel *two* points, then the touch sensors are no more than 2 millimeters apart. If only one point is felt, then put the second needle in succeeding holes until two points are felt by the subject. Make sure your subject does not watch.

Touch receptors can also distinguish textures. Test various parts of the skin for the ability to feel things that are rough, smooth, soft, and any other textures that you can think of.

Another investigation on sensors could focus on temperature receptors. For cold receptors, a wire is cooled on an ice cube and then placed on various parts of the skin. A paper clip that has been straightened out is suitable. Be sure to cool the wire after each application. For heat receptors, a wire is heated in very hot water before it is placed on various parts of the skin.

The amount of air the lungs can hold is measured with an instrument called a **spirometer** (*SPY roh MET er*). You can construct a simple spirometer out of 3 feet of ⅜ or ½ inch plastic tubing, a gallon jug, a dish pan, and a measuring cup. Mark the jug into 250 milliliter measures by continuing to add this amount of water and indicating each new level with tape. Fill the jug and dish pan with water. Turn the jug upside in the dish pan. Don't let any air into the jug. If any gets in, start again. Place the plastic tubing into the opening of the jug. Exhaled air that is blown into the jug will push an equal volume of water out. Use this simple spirometer to measure the lung capacity of your subjects. Measure ordinary breathing and deep exhaling. You might think of some feature that may vary among your subjects, such as adults who smoke and who don't smoke.

When you reach for a book on a high shelf, your brain has to make some decisions about how your body will move in order to get the book. Without any conscious thought, you stand on your toes, reach out your fingers, grab the book, and bring it down. To do all these movements, the brain must sort through much information from the eyes, the balance equipment in the ear, and muscle tension in the arm, hand, and fingers. This is said to require coordination. Eye-hand coordination can be tested with a homemade piece of equipment. The design is shown in the illustration. A person moves the wire loop along the coordination testing path created by the coiled wire. The purpose is to *not* touch the loop to the testing path. The circuit is designed such that the bell will ring if the wire loop touches the coiled wire. An investigation could be designed to see if there are differences between boys and girls, times of the day when the test is done, and people of different ages. You can probably think of other situations to investigate that may influence coordination.

You can test your coordination with this equipment. You can easily build it yourself.

These investigation science projects allow you to make predictions, something scientists want to do with the data they have collected. What is *most* important about the project you choose is that you are curious about it and involved with it.

☞ *Surveys*

Surveys are made by newspapers, TV stations, advertising companies, and people running for elected office to find out what the public thinks. Doctors and other health-care professionals conduct surveys to discover patterns of health and disease among various people.

A survey could be planned and carried out for a science fair project. Many topics are possible. Here are some possible survey topics:

- ❖ Seat belt and harness use. How many students in your school use a seat belt or harness when they ride in a car? How many families require that everyone in the car wear a seat belt? You might stand at an intersection where drivers slow down and count people who do and don't wear a seat harness. *If you survey traffic, make sure that you are in no danger from the moving cars. It is a good idea to have an adult see if your location is safe.*
- ❖ Weather and mood. Does the type of weather affect how people feel? Does a change in the weather affect how people feel? Does a change in the weather affect how people act? Ask your teachers about changes in the students' activities a day or so before a storm.
- ❖ Awareness of health risks faced by smokers.
- ❖ Awareness of health risks associated with drinking alcoholic beverages.
- ❖ Awareness of health risks associated with the typical American diet that has too many fats.

Be sure that your survey samples enough people. If you only survey a few people (10-15), your conclusions may not represent the population in general.

CHAPTER 3

Planning Your Project

Stating your purpose

When you have finally picked a problem to study about the human body, you then need to state the purpose of the experiment that you want to do. You can either make a statement or ask a question:

"The purpose of this project is to determine if body temperature changes in a regular pattern during waking hours."

or

"Does body temperature change in a regular pattern during waking hours?"

Notice that each of these sentences clearly states what the experimenter wants to find out. You should take some time to write a sentence that defines the problem you have decided to tackle. This will help make clear what you want to do during your investigation. The statement of your problem will be written in your *lab notebook*. Later in this chapter, you will learn more about this notebook and the "stuff" you should keep in it.

Gathering information for your project

☞ *The library*

The first place you want to visit to find information for your project is the library. This is where you will find books, encyclopedias, magazines, and newspapers with articles about the human body.

The card or on-line computer catalog will tell you what books the library has. Look up the subject headings "Medicine," "Human Body," and "Science Experiments."

To find magazine articles about the human body, use the *Readers' Guide to Periodical Literature*. There is a volume for each year. Look up the topic "Medicine," or the name of the part of the body you want to study, like "Heart" or "Eye." For each article you will find the magazine's name, the volume number, and the pages on which the article appears. Ask the librarian for help if you need it.

Scientists use libraries a lot to find out what is going on in their area of interest. So this library search is good experience for a budding scientist.

As you read, it is important that you take good notes. Index cards (3 x 5 inches) work well. Don't write on scraps of paper. They are easily lost. Write down information that will help you organize your ideas and to plan the steps you will take to do your project. Each index card should include the author's name (if any); the name of the book, magazine, or newspaper; the date it was published; and the pages you read. Later, you will rewrite important index card information into your lab notebook.

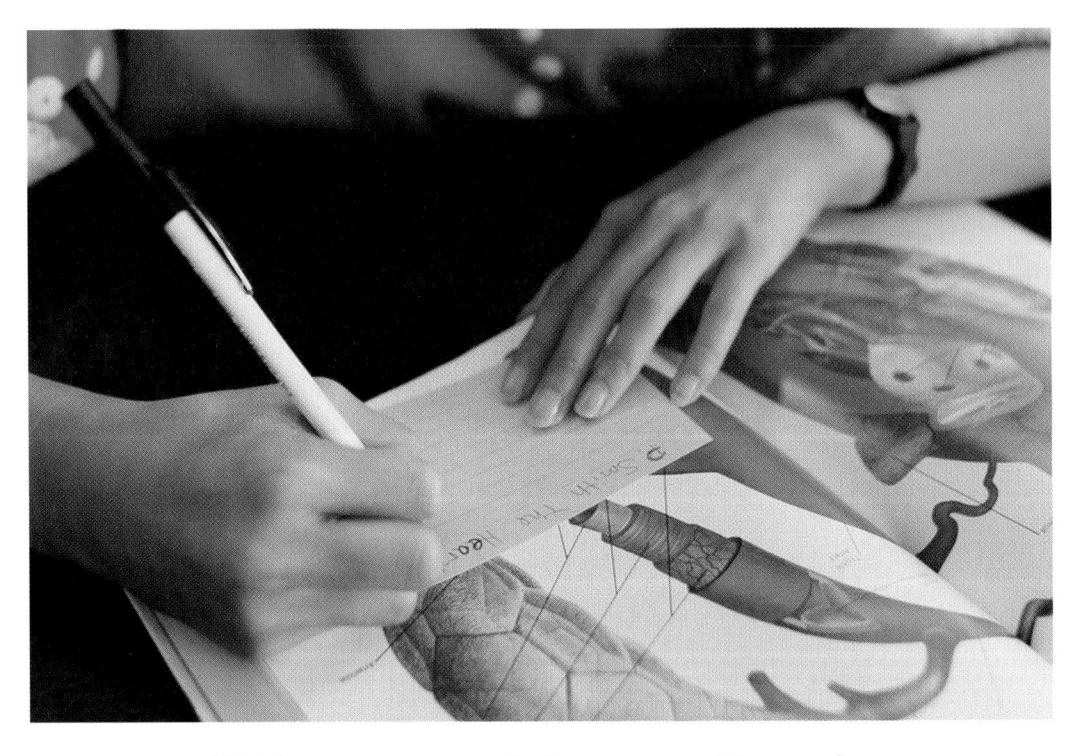

Writing out note cards is a smart thing to do.

Use your own words when you summarize what a book or an article says. **Plagiarism**, that is, copying word-for-word, or changing just a few words here and there, is not proper behavior for any scientist. Teachers and science fair judges will not accept it. What you write on your display and in your report about your project *must be in your own words*.

☞ *Resource people*

Your community probably has several people who can give you some help on your project about the human body. There are nurses, dentists, doctors, pharmacists, optometrists, and physical therapists. Perhaps there is a college nearby where you will find a helpful biology professor. Don't hesitate to get all of the information you can. The more you know, the better your project will be.

A dentist can tell you about teeth.

How an investigation is done

☞ *Experimental design*

Once you have decided *what* you want to find about the human body, the next step is to decide *how* you will do it. You must decide what materials you need to use and what you will do with them. This is called the **experimental design.** In a scientific investigation, the answers to your problem should come from the experiments you plan.

A well-designed science project that investigates a problem about the human body is often done under **controlled** conditions. This means that you intentionally change certain conditions to see how the part of the body you are examining changes. The conditions of an experiment that change are called the **variables.** You change the **independent variable** to see how the **dependent variable** changes. Here's another way of explaining these variables. The independent variable is something that you manipulate. The variable that responds to this manipulation is the dependent variable.

This may still be a little confusing. So, let's see how these terms apply to the experiment about the taste map that was described in Chapter 1. In this experiment solutions with different tastes are placed on the tongue. An independent variable (a variable that is changed on purpose) is a taste—sweet, salty, bitter, or sour. A dependent variable (a variable that responds to the change) is a place on the tongue that can detect a particular taste.

An experiment also needs to be controlled. This means that the subject must not be told what solution is being applied. After each application the mouth is rinsed out with cool water. The same amount of flavored solution is applied each time on the tongue. You might come up with some other conditions that must be the same for all subjects.

An experiment must have results that are repeatable. This means that you do the experiment again to make certain that you get the same results that you obtained the first time you did the experiment.

It's a good idea to ask your teacher, parent, or other knowledgeable adult to check over your design. Let them make changes that might make the experiments safer or more practical. However, make sure that the plan is *your* plan.

☞ *Lab Notebook*

You will need a lab notebook in which to write down everything about your experiment. This includes what you want to find out and what your guess, or hypothesis, is for the outcome of the experiments. The notebook also contains a list of the materials you use and how they are used. You record what you observe or measure. You write down what you think the data you collected means. You list all of the sources you used to gather information to help you in the project. This includes people, books, and magazines. The first page will have the project's title, followed by your name, grade, and school. A table of contents at the beginning will help readers find the different parts of the project.

Everything you do in an investigation is written down in a lab notebook.

The lab notebook could be a spiral-bound notebook or a three-ring binder. It should be used only for your science project. Remember to write everything down in this book—don't use scraps of paper. The notebook does not have to be real neat, but your handwriting should be clear enough so other people can read your ideas. Your teacher and science fair judges will look at it.

☞ *Materials*

The things that you will use for your project on the human body must be chosen with care. The list of materials must be *exact*. It states *what* and *how much* will be needed to do the project.

You should have little problem finding materials to be used in your project on the human body. In the section that gives project ideas, some materials for making equipment are listed. Other sources are listed at the end of this book. If you want to buy something from a biological supply company, ask your teacher to place the order. Remember that it may take several weeks for the order to reach you, so begin to plan your work early.

☞ *Procedures*

The steps that you will follow in your experiment must be carefully organized. As you think about how you will do your experiment, write down each of the steps in the order they will be done. Include as much detail as possible. Your experimental design, or plan, should be written in your lab notebook.

A **flow chart** will help to remind you of the various steps that you need to follow in your experiments. This chart is a simple version of your experimental design. It reminds you what must be done next. Check off each step as you complete it.

☞ *Measure in Metrics*

All of your measurements should use the metric system. Scientists all over the world measure length, volume, mass, and temperature in metric units. Measuring in metrics is easy because it is based on the number 10. This is just like American money. One dollar can be divided into 10 dimes or 100 pennies. In the metric system, you will not have to worry about fractions. You will use decimals.

The basic metric unit for length is the **meter** (m). One meter is a

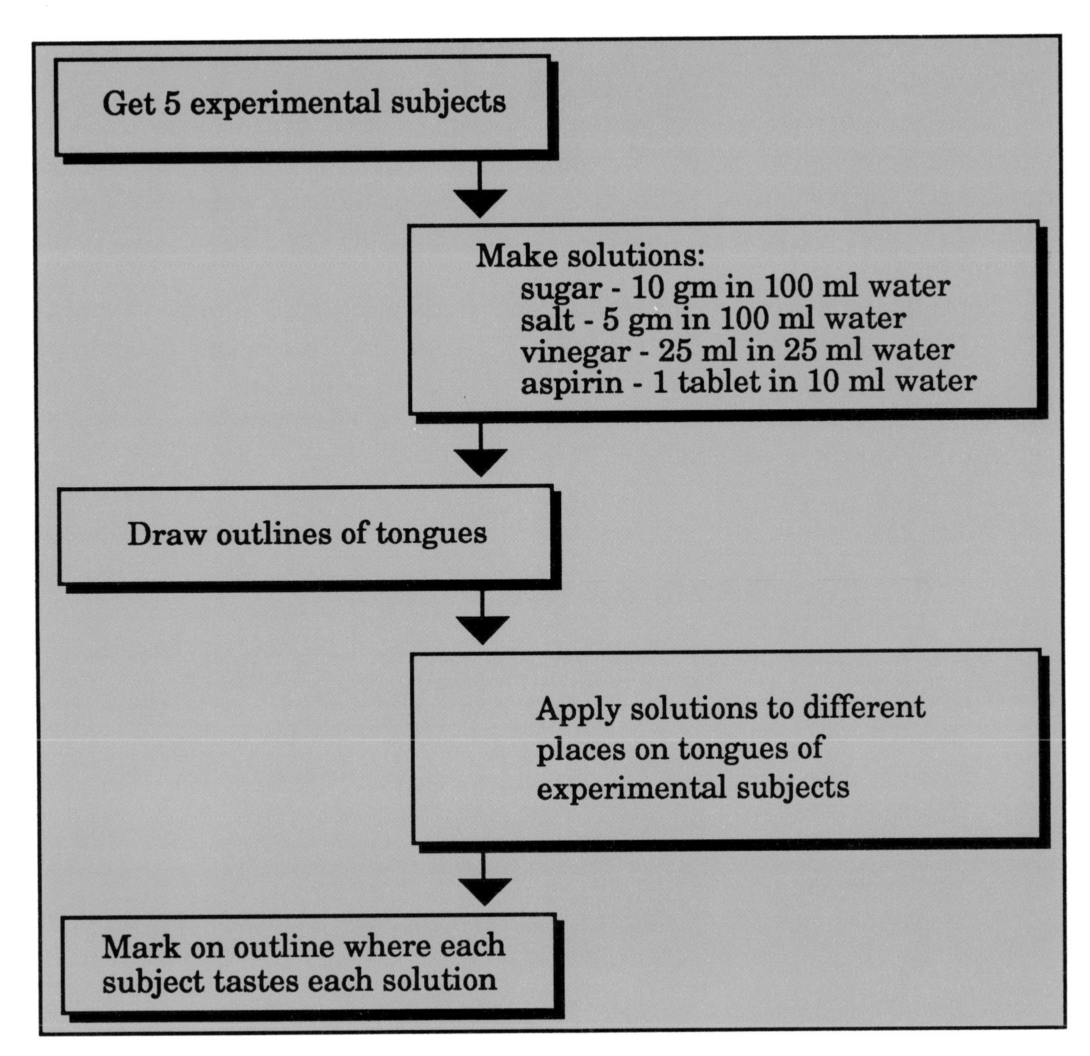

This is a flow chart for the experiment on the tongue and taste.

little longer than one yard. A 100-yard football field measures 91.5 m. A meter is divided into 100 equal and smaller units called **centimeters** (cm). A meter is also divided into 1000 equal and still smaller units called **millimeters** (mm). Most rulers will have the metric scale printed on one side.

The basic metric unit of volume is the **liter** (l). One liter of lemonade is a little larger than one quart. A liter is divided into 1000 equal and smaller units called **milliliters** (ml). A 12 ounce can of soda pop holds 296 ml. Measuring cups and medicine droppers can be used to measure amounts of liquids in milliliters. Your school might have graduated cylinders for measuring liquids.

The basic metric unit of **mass** is the **gram** (g). Mass measures the amount of matter, or "stuff," present in an object. Many people confuse mass with weight. Mass is not influenced by gravity as weight is. Large objects have their mass measured in **kilograms** (kg). Some electronic kitchen scales will measure grams.

Temperature in the metric system is measured in degrees Celsius (°C). A nurse uses the Fahrenheit (°F) scale to determine your body temperature. Water freezes at 0°C, or 32°F, and boils at 100°C, or 212°F. Body temperature is 37°C, or 98.6°F. A thermometer on a nice spring day might read 25°C, or 77°F.

Metric System Symbols

Length	Volume	Mass
meter = m	liter = l	gram = g
centimeter = cm	milliliter = ml	kilogram = kg
millimeter = mm		

☞ ***Time schedule***

You should begin to plan your project several months before it is due. This will give you enough time to gather materials, do the project work, analyze the information you gather, write a report, and prepare a display. Estimate how much time each step should take and then add a little more time.

Here is a sample time schedule for an investigative project:

Week 1	- Choose a problem.
Weeks 2 and 3	- Gather information.
Weeks 4 and 5	- Plan your experiment and gather your materials.
Week 6	- Begin your experiment.
Week 7	- Continue to collect data from your experiment.
Week 8	- Finish data collection; begin to analyze data.
Week 9	- Prepare graphs; make display.
Week 10	- Practice oral report; science fair starts or project is due.

You will have fewer hassles if you carefully plan a time schedule. This will help make sure that your project is finished in plenty of time before the science fair.

CHAPTER 4

Doing Your Project

Safety

"Safety first!" is the rule when working on a science project. Here are some tips to keep you from possibly being injured. A responsible adult should check your project plan for possible hazards.

If your project uses human subjects, they must be informed of what will be done with them. They must agree to serve as experimental subjects. Have your experimental design checked over by the school nurse or by a physician if the experiment includes physical activity or exposure to chemicals by the experimental subjects.

It's a good idea to go over your experimental design with an adult.

Have a place of your own to work on your project. It might be a table in your classroom that your teacher assigns to you. The kitchen, dining room or your room at home are also possibilities. You will need a desk to write on. Keep your work area neat. As you do your project, be sure to clean up. Put away things that you no longer need.

If any chemicals are used in your project, use them *only* with an adult to supervise you. Wear protective glasses, rubber gloves, and a rubber apron. Wash your hands each time you finish using the chemicals. Keep your hands away from your mouth while you work. All chemicals should be clearly labelled and stored in a suitable container.

You should use batteries to power any electrical equipment. It is too dangerous to use electricity from a wall socket. Even though the current available from batteries is not strong, wires coming from them can become hot and thus possibly cause a burn.

Avoid using sharp objects that can puncture the skin.

Keeping a record

The lab notebook in which you wrote background information is also used to record your observations of the experiment. You can't write down too much! Don't expect to remember later what you saw or measured. Write down your measurements and observations *right away*!

There are two kinds of observations that you can record. A **qualitative** observation does not use measurements: "Fingertips had more touch sensors than an arm." A **quantitative** observation uses measurements and numbers: "The touch receptors on fingertips were spaced closer than two millimeters apart, while those on the arm were at least 10 millimeters apart." Quantitative observations are more *precise*. Whenever possible, use quantitative observations for your project.

A scientist must be **unbiased** when observing what happens in the experiment. This means that you don't make the results come out the way you want them to. If the results are not turning out as you predicted, that's OK. Maybe your hypothesis is not correct. That's OK, too. An incorrect hypothesis means that you can rule out your "educated guess" as a possibility. Honesty is expected of scientists.

When you compare measurements of different groups, you will want to figure an *average* measurement for the subjects in each group. It's a way to represent the measurements for all subjects within a group.

The arithmetic you use to compute an average is simple to do. Let's suppose that you are measuring the pulse rates of subjects while they are lying down. These pulse rates are added together. The sum is then divided by the number of subjects in this experimental group. The answer from this last calculation is the average measurement for the group.

Pulse	72		
rates,	68	70	The average pulse rate
beats/minute	76	5	350	for fifth grade boys
	70	35	while they are lying
	64	0	down is
	350	0	**70 beats/minute.**

Photos of you as you do different parts of your investigation are one way to keep a record of your observations. These pictures should be clear. Make sure that your subjects are well-lighted. Hand-drawn illustrations that are neat can also be used.

Coming to conclusions

When you have completed your experiments, it is time to come to conclusions. You did the experiment to get an answer to a question. Did your observations support your hypothesis? Explain everything in your lab notebook. If your data don't support your hypothesis, don't worry. Avoid stating that the experiment was a "failure", or that it didn't work. Just explain what happened. Maybe some variable was not controlled. Go back over your notes in the lab notebook to find out where an mistake might have been made.

Most important in your conclusions is a statement of what your experiment means. For example, suppose you test people who are very tired on the coordination machine. Your data show that they make more mistakes, that is, touch the wire more often than when they are rested. You could then discuss why people who work with dangerous or expensive machinery should get enough sleep. Thus, your conclusions tie together what you found out from your experiment and the world in which you live.

CHAPTER 5

Presenting Your Project

Making graphs to show your data

Numbers you collected from your observations can be formed into **graphs**. Graphs will pack lots of information into a little space. They are a good way to present your data because they make it easier to understand the information. Paper on which to draw graphs can be bought at office supply departments. It comes with lines printed on the sheet. Use graph paper that has four or five lines per inch.

A **bar graph** can be made when you want to compare several groups. Bars may be drawn to represent the average amount measured for each group. A bar graph would be used to compare the numbers of individuals observed to have differing traits. To see how a bar graph can be made, let's use some data collected from an experiment. It shows the number of minutes it took for the pulse rates of boys and girls who hopped for 60 seconds to return to normal.

Groups	Average time, seconds
Fourth grade girls	180
Fourth grade boys	205
Fifth grade girls	240
Fifth grade boys	220
Sixth grade girls	195
Sixth grade boys	200

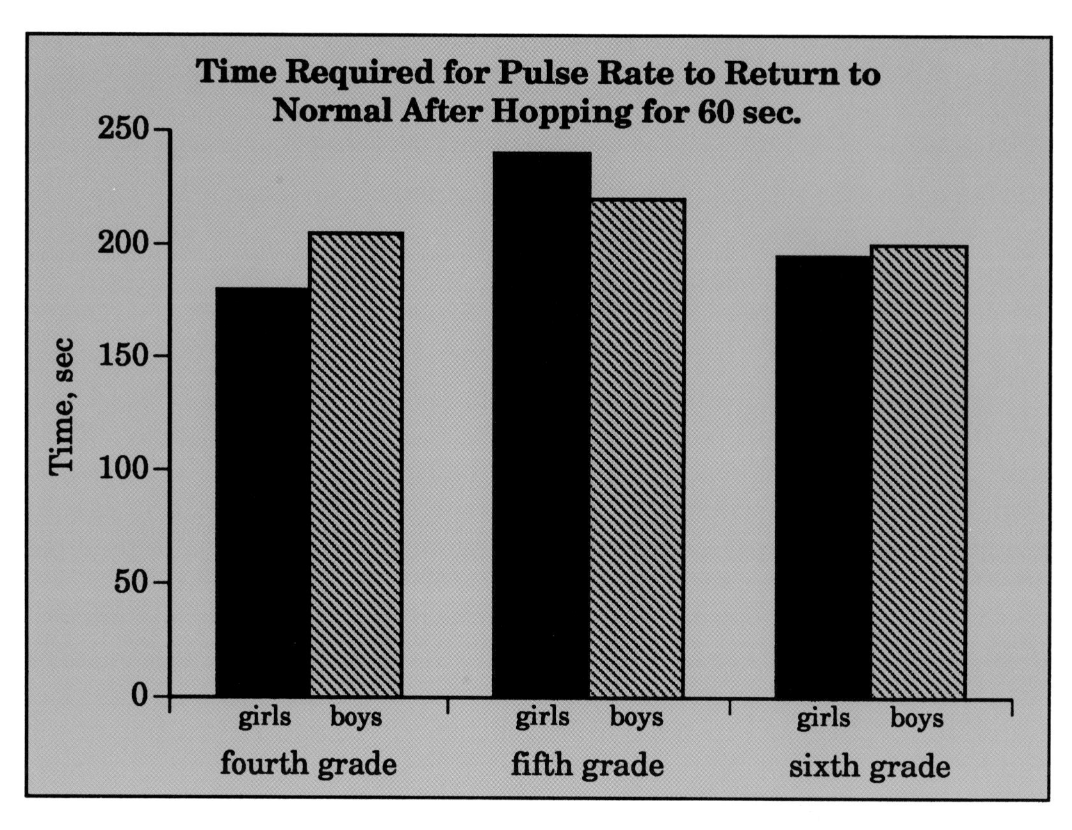

A bar graph compare data taken from groups.

A line (the **horizontal axis**) is drawn at the bottom of the graph. This baseline has a place for each group. This line represents the independent variables, or the variables that you deliberately change. For our example, the independent variable is the gender; the gender have been intentionally chosen. The line on the graph's left side (the **vertical axis**) displays the amount of each dependent variable, or the measurement that changes in response to the change you cause. The average time for the pulse rate to return to normal after vigorous exercise for each group is the dependent variable in our example.

Drawing a bar graph to show your data takes careful planning. A bar's height is proportional to an amount. Thus, you must decide how high each bar will be drawn on the graph paper. In our example, one cm on the graph paper represents 50 heartbeats/minutes Choose how wide the bars will be—the bar widths must be the same. The bars can be drawn in different colors to make each group stand out.

A **pictograph** is constructed like a bar graph except that it uses a repeating symbol to represent the amount of an item. If there is a fraction in this amount, only a part of the symbol is used. The symbol you choose to use to represent an amount should have a simple shape that readers can easily recognize. As for a bar graph, you should plan a pictograph's layout. Decide how many symbols will need to appear in each row. Because you will use many copies of this symbol, all of them in the pictograph must look the same. There are several ways to do this. You could carefully draw your design and then make many copies on a copy machine. You could, instead, cut out your symbol from a piece of cardboard and then trace around the design.

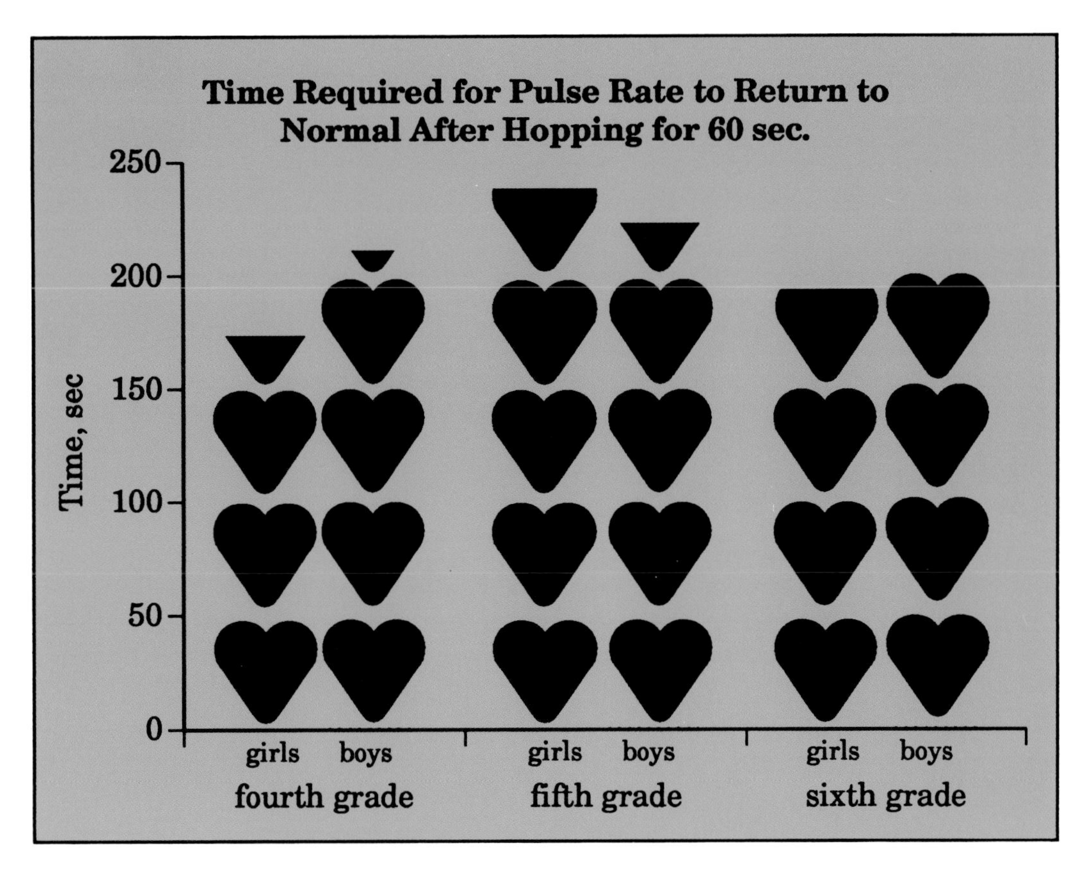

A pictograph is a type of bar graph.

A **line graph** uses a continuous line to show that the dependent variable in an experiment changes as the independent variable is changed. Frequently, a line graph shows how an experimental group changes over time. In these situations, time (minutes, days, weeks) is the independent variable, and so it is plotted along the horizontal axis. If you prepare a line graph that compares data taken from different experimental groups, the line for each group can be a different color.

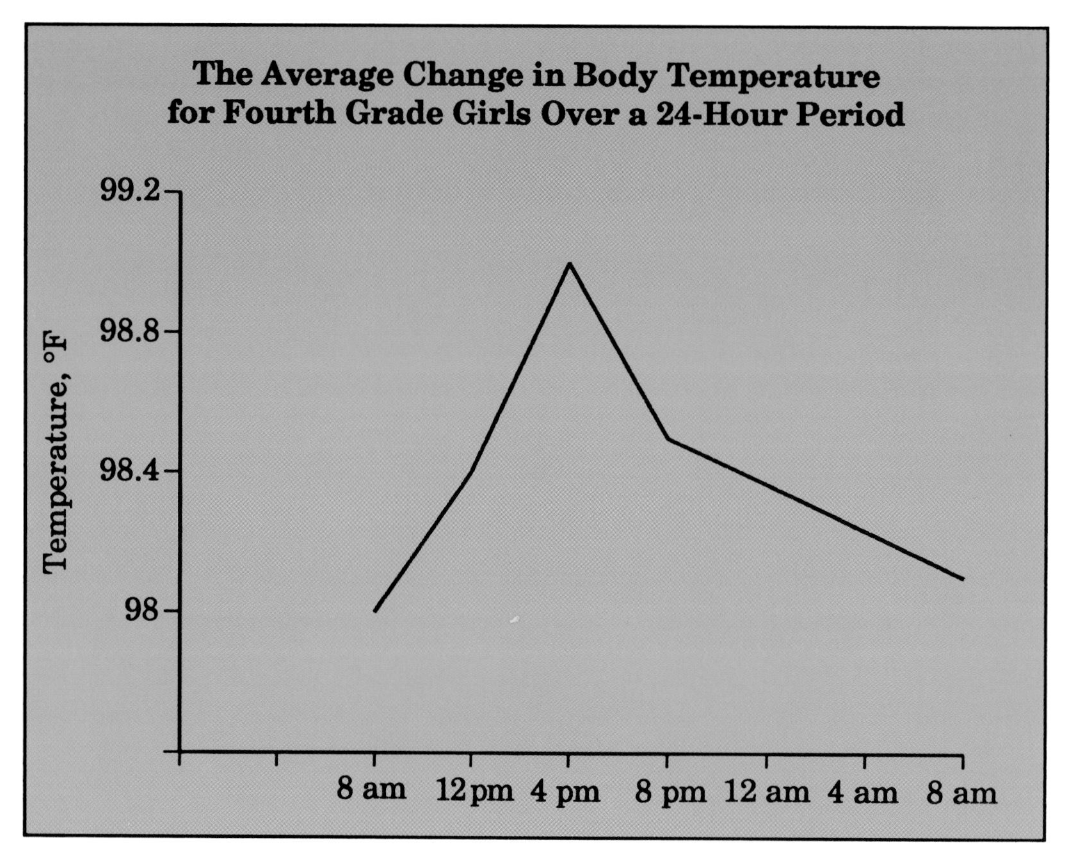

A line graph shows a change over time.

Each graph must have labels on the left side and at the bottom. These labels are written so that the reader will know what the numbers represent. In addition, each graph must have a title that tells what observations are being shown.

Lettering in a graph must be neat. You can use press-on letters purchased from an office supply store. Letters could be traced using a template. If you have access to a computer, it can be used to prepare a graph with proper labels.

Putting your display together

Visitors to your science project must be able to quickly understand what your project is about and what you did. A science project display tells a short story of your work on the project. It should show:

- ☆ A statement of the problem—what you wanted to find out.
- ☆ The hypothesis—how you guessed the experiment might turn out.
- ☆ Materials and methods—what materials you used and what you did with them.
- ☆ The results—what you observed.
- ☆ Your conclusions—did you find out what you wanted to know?

Science fairs usually have rules on the basic display design. Check with your teacher for local regulations, especially on the size of the space made available for you to show your project.

The most common display has three panels and stands by itself. Several sturdy materials are available for these panels. Panels can be made of ¼ or ⅜ inch plywood held together with hinges. The plywood can be painted or covered with cloth or paper. Foam board is lightweight, but strong enough to hold your written materials. Panels are easily cut out with a knife. Cloth tape can be used to make hinges that hold panels together. Foam board is available at office supply stores. Pegboard is also a handy material to use. It has holes already in it for hanging items. Because pegboard may bend if heavy objects are attached, you can nail 1" x 2" wood strips around the edges of the pegboard. Once these wood strips have been nailed on, hinges can be added. Some office supply stores and biological supply companies sell cardboard displays.

The center panel shows the project's title and some of the graphs you made from the data. Pictures of the experimental setup could be placed here instead of the graphs. On the left side, the statement of the problem and your hypothesis are shown. The right side can display any other results, as well as your conclusions. Each of these items is written or typed on a separate piece of paper.

Letters for the title can be cut from construction paper and should be at least two inches high. While neat hand printing is OK, your display will look better if you use a typewriter or word-processor for the other lettering. Since visitors to your display will stand several feet away, letters should be at least ¼ inches high. Be sure to have someone check the spelling, grammar, and punctuation in your display before you put it all together.

It is important to make a neat display.

You might want to include drawings or photographs if they help explain your investigation. These illustrations need to be labelled.

How you attach materials to the display board is important. Staples look cheap! Tape hidden behind larger pieces is good. Test glues to find one that does not wrinkle paper after it dries. Before you fasten the pieces in place, lay everything out to see how it will look and fit on the display board.

Your written report

In addition to the lab notebook that was discussed in Chapters 3 and 4, some science fairs will expect that you prepare a written report on your project. This formally presents your work on the project. It shows readers the details of your efforts.

If possible, use a typewriter or computer. However, *very* neat handwriting is acceptable. The pages can be bound into a folder or a binder.

On the first page is the title of the project, your name, grade, and school. The table of contents is the second page. Readers can more easily find different sections if they know what pages to look up. The statement of purpose explains why you wanted to work on this project about plants and what you wanted to find out about them. Included in the background research section are summaries of the articles and books you may have read (remember to use the note cards you made), the names and titles of the people to whom you may have talked, what you learned from them, and any other materials you found to help you plan your project.

A Tongue
Taste Map

Keesha Martin
Grade 4
Martin Luther King School
Uniontown, Virginia

Table of Contents

Statement of Problem........3
Background Research.........4
Hypothesis...........................6
Experimental Design..........7
Materials..........................8
Procedures.......................9
Data (Observations and
Measurements...............11
Conclusions.......................13
Bibliography......................15
Acknowledgments............16

The materials you used, how you used them, what data you

obtained, and what the data mean are the next sections. The information you should include in these sections has been discussed earlier in this book.

The bibliography lists all articles and books you read. These materials are listed alphabetically by the authors' last names. Here are some examples to help you prepare a bibliography:

For a magazine:

Barinaga, Marcia. "The secret of saltiness." Science 254:664-665 (1991).

For a book:

Hodges, Laurent. Environmental pollution. New York: Holt, Rinehart and Winston, 1977.

Finally, you need to acknowledge all of those people who helped you on the project, thanking first those who gave you the greatest support and help.

Judging of your project

You will probably be expected to talk about your project to science fair judges or to your class. In four or five minutes you will have to:

- Introduce yourself, giving your name, school, and grade.
- Give the title of your project.
- Give the purpose of your project.
- Tell why you choose this project.
- Explain what you did.
- Show your results—explain any graphs or pictures that are in your display.
- Give your conclusions if you did an investigation project.
- Explain what you learned.
- Ask if there are any questions.

You can use notes to help remind you of what to say. However, if you practice several times, you won't have to look at them too often. Practice your talk while a parent, other relative, or a teacher listens. Ask this person for helpful suggestions.

When you finally give your talk, stand to the side of your display so the judge or other viewers can see your work. Talk slowly, even if you are really nervous! If you don't know the answer to a question, be sure to say you don't know. Remember, honesty is important.

The science fair judges who visit with at your display will evaluate your project for an award. Each science fair usually prepares its own judging sheet. However, most will score projects in these areas:

✔ Scientific thought - The judges will see if your project follows the scientific method. Has the problem been clearly stated? Are the procedures proper and thorough? Have controls been properly used?

✔ Creative ability - The judges will want to know how you chose this topic. Your score will be lower if you repeated something you read in a book or if someone else did the actual work. If a book, like this one, gave you the idea for your project, how did you use your imagination to develop the project more fully? More points will usually be given for scientific thought and for creativity than for the other areas.

✔ Understanding - Judges will ask some questions to see if you understand the key scientific features of your project. If you have prepared an exhibit, does the display provide some answers to questions about the topic?

✔ Clarity - The judges will examine your project to see if it clearly presents the hypothesis, procedures, data, and conclusions. Will the average person understand the project?

✔ Technical skill - Finally, the judges will check on how your display appears. Did you do most, if not all of the work? How attractive is your display? Did you carefully check written materials for correct spelling and grammar?

Final words of encouragement

When you have presented your science fair project to your family, friends, and judges, feel good about what you have done. You have worked hard. You have become more aware of what scientists do. Science is "doing," not just memorizing some facts. You have learned how to find answers to a question about plants. By doing this project, you have discovered more of the wonder of science. Get fired-up and do another science fair project next year!

A good project gets recognized.

WHERE YOU CAN BUY SUPPLIES FOR EXPERIMENTS ABOUT THE HUMAN BODY

Carolina Biological Supply Co.
2700 York Road
Burlington, NC 27215
1-800-334-5551 (East of the Rockies)
1-800-547-1733 (Rockies and West)
1-800-632-1231 (North Carolina)

Connecticut Valley Biological Supply Co., Inc.
P.O. Box 326
Southampton, MA 01073
1-800-628-7748 (U.S.)
1-800-282-7757 (Mass.)

Nasco
P.O. Box 901
Fort Atkinson, WI 53538-0901
1-800-558-9595

Science Kit & Boreal Laboratories
777 East Park Drive
Tonawanda, NY 14150-6784
OR
P.O. Box 2726
Santa Fe Springs, CA 90670-4490
1-800-828-7777

Ward's
P.O. Box 92912
Rochester, NY 14692-9012
1-800-962-2660

GLOSSARY

artery - a blood vessel that carries blood from the heart to other parts of the body.

bar graph - a graph that uses columns to compare different values obtained for experimental groups. The height of each bar is proportional to the value.

blood pressure - when blood is pumped out of the heart, the force pushes against arteries.

centimeter - the distance measured by 1/100 (0.01) meter. There are 2.54 centimeters in one inch (abbreviation: cm).

conclusions - what you interpret the results of an experiment to mean.

control group - a group in an experiment in which as many variables as possible are kept constant because they could affect the outcome of the experiment.

data - the observations and measurements that you make in an experiment.

dependent variable - the factor or condition that changes as a result of the presence of, or a deliberate change you make in, the independent variable.

diabetes - a disease in which the blood has too much sugar in it because there is not enough of the hormone insulin to help sugar get into the body.

diastolic blood pressure - the pressure of blood in arteries when the heart is resting.

experimental design - the plans you make so you can do an experiment. The design includes what you will use and how you intend to use them.

experimental group - a group in which all variables are the same as those in the control group *except* for the factor that you are following in your experiment.

flow chart - a list that is a shortened version of the steps you want to follow in doing your experiment. As you complete each step, you should check it off the list.

gram - the basic unit of mass in the metric system. There are 28.3 grams in one ounce (abbreviation: g).

hormone - a substance released into the blood by a body organ and acts as a messenger to another part of the body. Hormones affect body growth and development, the ability to use certain foods, and to react to events that happen to us.

hypothesis - a statement that gives a possible answer to a question. Because you may already know something about the question, a hypothesis is sometimes called an "educated guess." To see if it is true or not, a hypothesis is tested by doing an experiment.

horizontal axis - the line on a graph that goes across the bottom. It is used for showing values for the independent variable.

independent variable - the factor or condition that you want to study. In an experiment, you intentionally change this factor.

insulin - a hormone that regulates the amount of sugar in the blood.

kilogram - the mass of 1000 grams. One kilogram is equivalent to about 2.2 pounds (abbreviation: kg).

line graph - a graph that uses a line that shows the dependent variable changes as the independent variable is changed.

liter - the basic unit of volume in the metric system. One liter is a little smaller than one quart (abbreviation: l).

mass - the amount of matter, or "stuff," that is present. Weight is often confused with mass. Weight is the pulling force of gravity on matter.

meter - the basic unit of length in the metric system. One meter is a little longer (39.4 inches) than one yard (abbreviation: m).

milliliter - 1/1000 (one one-thousandths) of a liter. There are approximately 28 milliliters in one fluid ounce (abbreviation: ml).

millimeter - 1/1000 (one one-thousandths) of a meter. There are approximately 25 millimeters in one inch (abbreviation: mm).

organ - a part of the body that does a particular activity. For example, the heart is the organ that pumps blood through the body.

pictograph - a graph that uses a series of pictures to show the values measured or observed for the dependent variables. It is put together like a bar graph.

plagiarism - copying word-for-word what someone else has written and not giving credit to that person.

prediction - what you think will happen in an experiment.

qualitative observation - an outcome of an experiment that is not an amount that can be measured, such as color.

quantitative observation - an outcome of an experiment that is measurable, such as numbers of individuals.

receptor - a structure on an organ that receives a signal or stimulus that is carried by nerves to the brain.

results - what you measure or observe as an experiment is carried out.

scientific method - a systematic strategy scientists use to discover answers to questions about the world. It includes making a hypothesis, testing the hypothesis with experiments, collecting and analyzing the results, and arriving at a conclusion.

sense organ - an organ that picks up signals from outside the body and converts them so that nerves can carry the information to the brain.

sphygmomanometer - a cuff that is wrapped around the arm when blood pressure is measured.

spirometer - an instrument that measures the volume of air in the lungs.

stimulus - a signal, such as light or sound, that the body can detect with a sense organ.

systolic blood pressure - the pressure of blood pushing against artery walls as blood is pumped out of the heart.

taste buds - structures on the tongue that have receptors for converting a taste stimulus into a signal that nerves carry to the brain.

unbiased - not allowing your preferences to interfere with collecting or analyzing data in an experiment.

variable - some factor in an experiment that can be changed.

vertical axis - the line on the left side of a graph. It shows the values of the dependent variables.

INDEX